Dreamstar

**It's not about being right,
it's about thinking outside the box.**

Damon Dion Reed

authorHOUSE®

AuthorHouse™
1663 Liberty Drive, Suite 200
Bloomington, IN 47403
www.authorhouse.com
Phone: 1-800-839-8640

This book is a work of non-fiction. Unless otherwise noted, the author and the publisher make no explicit guarantees as to the accuracy of the information contained in this book and in some cases, names of people and places have been altered to protect their privacy.

First published by AuthorHouse 7/10/2007

ISBN: 978-1-4259-9672-7 (sc)

Library of Congress Control Number: 2007901172

Printed in the United States of America
Bloomington, Indiana

This book is printed on acid-free paper.

I would like to dedicate this book to my parents who guided my growth, maturation, and love for society: from then to there. I would also like to dedicate this book to Zeus's parents. They have been invaluable in my understanding of life. They have put up with my quirkiness through thick and thin, even though I often seem like a crazy person. I would also like to thank my brothers. Without them, I would have never strived to beat them intellectually, even though I often seemed to lurk in the shadows of their memorizable intelligence.

Contents

List of Figures

Chapter One
Big bang? Not!

How we think depends on what we have learned or experienced in our lifetime. Our knowledge creates a mental box that guides our perception of the world and its problems. This idea is NOT novel. Therefore, I think as a biologist/chemist. With that said, I have decided to tackle astronomy and quantum mechanics by thinking as a biologist/chemist. These ideas are NOT SUPPORTED by ANY SCIENTIFIC DATA. By asking "WHAT IF" and then thinking about the world through the mind's eye of a biologist/chemist, the current big bang theory just doesn't make any sense.

If the big bang theory is true, then all the matter/energy in the universe was in one place at one time. IF everything was in one place and it exploded, then it would disperse the energy/matter in a spherical orientation and you could pinpoint the center of the universe. Problem number one: There is no center to the universe. Ask an astronomer where the center of the universe is located and they will probably laugh at you. The mere fact that we are living on a planet of heavier elements is proof that the big bang didn't happen the way that theorists have described it. Now many of you might be wondering how I came to that conclusion, so here is my stupid reasoning. If a big bang created all this matter, which dispersed to form galaxies, then the only thing there was to begin with was… PROTONS!!! Now since we know that YOUNG stars make elements like hydrogen or helium and OLD stars, i.e., supernovas, make the heavier elements, then the planet we exist on, which is made from heavier elements, must have resulted from the death of millions of stars. Ok, I'll go a little more slowly. Since dying stars make heavier elements like iron, lead, carbon, and oxygen, then the planet we exist on was formed from the remnants of dead stars, i.e., heavier elements. Now since we are in a galaxy of relatively young stars and the closest neighboring galaxy is about 2.5 million light years away (http://en.wikipedia.org/wiki/Andromeda_Galaxy), then where did the remnants of these dead stars come from? It would seem logical that another galaxy existed near where our galaxy exists OR another galaxy existed in the exact same spot where our galaxy is NOW. If our galaxy were to have occurred from a big bang, then where did all the heavier elements come from? Now there will be people that say these heavier elements occurred at some point in the big bang "process." All I have to say to that is "Abracadabra!" because it must have been magic. We see young stars make matter like hydrogen and helium, but only dying stars make heavier elements that make up this planet, other planets, and YOU & I, which means that everything we SEE and FEEL is made from remnants of dead stars. This means that something existed before our galaxy was born which made all these heavier elements!!!!

Now the big bang was postulated because distant galaxies are moving away from us at a faster rate than closer galaxies. This is called the Hubble Constant and scientists still can't agree on its value. (Earth*A short history of nearly everything, p 168)

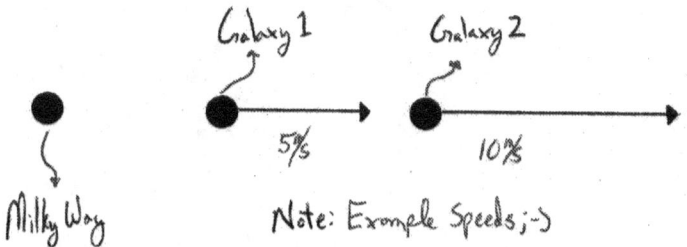

Figure 1: Galactic movement.

Since I can't argue with astronomical observations, I would just like to ask one question; "Is the big bang theory the ONLY plausible explanation for these astronomical observations?" To answer this, I need to look at a couple big bang scenarios.

If the Milky Way exists on an arbitrary "X" axis and the big bang was a uniform explosion, then no two galaxies should be moving in reference to the plane of the Milky Way at the same rate or in the same direction. Therefore every galaxy should have a unique galactic vector in reference to the Milky Way such that the vectors will point to one region of space where the big bang occurred. Even if the big bang was non-uniform, the resulting galactic vectors of each galaxy should still be pointing to a centralized region of space where the big bang occurred.

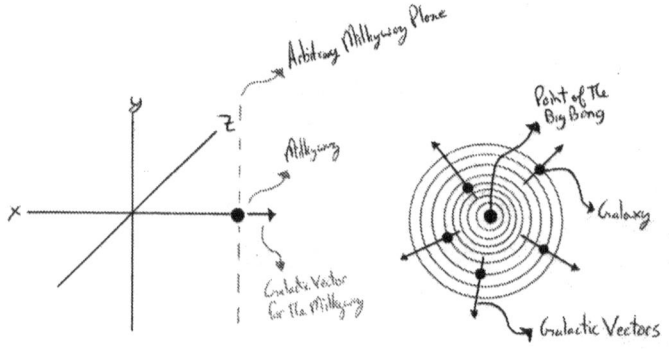

Note: If the Big Bang did happen, then NO two galaxies should have the same Galactic Vector!

Figure 2: Uniform big bang.

Therefore, IF the galactic vectors of each galaxy, with reference to the arbitrary Milky Way plane, are not pointing in one general region, i.e., seem random, then this disproves the whole big bang theory. Which leads us back to the question if there is another explanation to why "distant" galaxies are moving away from us at a faster rate? Now I have a stupid idea why distant galaxies might be moving away from us at a faster rate, but for now let's overlook the fact that the universe is not spherically orientated around one point and assume that all matter/energy was in one place at one time and exploded. BANG! Now this must have happened a really really long time ago because galaxies are really really far apart. The other possibility is that the energy that creates galaxies travels through bent regions of space-time, but I am going to have a tough time convincing people of that idea. So let us assume that it has taken a long time for galaxies to form and disperse across the universe, which means that all galaxies should be the same age. Problem number two: All the galaxies are not the same age. There are lots of young galaxies, old galaxies, teenager galaxies, thirty something galaxies, and grandparent galaxies. So lets overlook the fact the big bang created a universe in which some galaxies matured slower, some matured faster and some are still being born. Since galaxies mature at different rates, then this "big bang" was a dynamic event in which many variables don't add up under the current theory. So let's keep overlooking these inconsistencies and imagine that all this matter/energy is spreading out from one point, non-uniformly, to form galaxies which are aging at different rates. Now SOMEHOW a big ball of energy/matter that was the beginning of the Milky Way dispersed itself into even smaller units, i.e., stars, which started spinning in such a way that it created a center. Now the Milky Way is kind of like all those "other" spiral galaxies in the universe. Problem three: Numerous "similar" galaxies were formed from this big bang. Now just imagine a cherry bomb blowing up. Pop! Now let's think about how many "similar" fragments of the cherry bomb you will be able to find. I'm assuming that someone could find lots of small shreds and a couple large chucks of the cherry bomb, but the statistical chance of having numerous similar pieces of a cherry bomb spinning the "exact same way" through the space around where the cherry bomb blew up, is relatively small. If you were to correlate that idea to the big bang, then the big bang makes even less sense. So IF the big bang happened, then one small amount of energy from the big bang formed the Milky Way and this energy/matter organized itself into a spiral form, for whatever random reason, and now our galaxy started to age differently from all the other galaxies in the universe. The Sun in our solar system formed by the collection of what is thought to be matter and ignited into a big ball of fire. As the result of gravity, the Sun attracted all this matter, which was formed from the remnants of dead stars, to form planets. The Sun's energy then enabled our planet, Earth, to mature/cool at such a rate that it has spawned US. Yeah!!! Ok, I'm not really that excited since my life FEELS like it really really sucks sometimes, but I'm sure there are some rich bitches out there that are truly enjoying their lives. Next time you can't find anything on TV, go outside, look up, and ponder if all of this happened by chance OR if there was some order to it. Personally, I see massive amounts of order and this is why I think galaxies, stars, and people are the result of highly ordered energy. Now most of you are probably thinking "Oh Poop! Not another religious book!" My reply is of course "Heck NO this isn't a religious book! This book is about energy, order, and where did our galaxy come from?" I'll save my religious views for the end of the book;-) Now you might be getting the feeling that I'm crazy, so I'll tell you right now "I'm crazy!" So if you didn't think I was crazy before, then this next stupid idea is going to convince you that I am crazy.

My mother met my father at a fair in Wisconsin and a little magic happen, which some call fate. In short, they agreed to disagree and to have children. Now my parents had to grow together, I had to grow in my mother's womb, I was born, then I grew up physically by eating other things that have GROWN on this earth. So why do we still think that the every galaxy was formed as the result of one event? It doesn't make sense to me because most of what we SEE or FEEL has GROWN from something else! To put it simply, everything seems to be born, matures, gives birth to similar entities, and then dies to make room for new entities. Now I can't imagine the energy source that created life, worlds, galaxies and everything, but IF a SYSTEM that creates things which have a maturation pattern of being born, growing, giving birth to new things, and dying, then more than likely the SYSTEM goes through a similar process. The current big bang theory is analogous to the idea that you could plant an apple seed and it will grow into a banana tree. I just doesn't happen. Now the leading thought on the creation of the universe is that everything was formed by one event. Every Galaxy, every Planet, every Star was the result of one event. Now this is probably true, but it seems MORE logical that this "event" produced an entity that grew, matured, and gave birth to similar galaxies. So my first stupid idea is that something happened a long time ago and that EVENT formed something great, a galaxy if you may, which matured and gave birth to other galaxies that grew, matured and gave birth to more galaxies. I know it sounds crazy, but who came up with the big bang theory anyway? Also, how could they forget to correlate what happens in nature back to the SYSTEM that created us, i.e., our galaxy? Since we are a part of nature, and nature is created by matter which is subsequently formed by stars, then the true system that is a relative beginning to us and dictates all the laws by which we abide was passed down from galaxies to stars to matter to planets to nature to us. Now IF galaxies proliferate, how and why do galaxies give birth to galactic seeds so far away?

Chapter Two

Anybody care for a space-time galactic cone? Mmmm, yummy!!!

It is obvious that galaxies are really far apart. Therefore it is safe to assume that IF a galaxy gives birth to new galaxy, then there must be an explanation of why the birth of a new galaxy is so far away from the parent galaxy. The first and obvious reason that galaxies are isolated would be that a parent galaxy gives birth to an infant galaxy that has plenty of room to grow. I imagine that a galaxy propagates like a dandelion that goes to seed. One dandelion matures until the wind sends the fluffy white seeds floating gently into the warm after-noon breeze to find fertile ground where a new dandelion has ROOM to grow into something spectacular. (Yeah, dandelions spectacular? I sound like I'm on crack.) Like this, the parent galaxy probably produces prodigy far far away because the galactic seed needs lots of space to grow. Now if the galactic seed has a chance to mature into a "STABLE" spiral galaxy, then it could produce an environment that is suitable to support life, i.e., things like You and I. As for how a galaxy produces prodigy far far away, well it might have something to do with that super-massive black hole that seems to be at the center of spiral galaxies.

A super-massive black hole can manipulate space-time and has a tendency to suck in a lot of energy in the form of matter, stars and light. So I think it is a prime candidate for being able to give birth to a new galaxy far far away. First thing I would like to point out about a black hole is that we don't know what a black hole "is." (A Brief History of Black Holes, Progress in Physics, April, 2006, Vol. 2, p 54) It might just be a black hole or as I like to think of it, an Energy Sink or Birthing Canal;-). Energy enters the black hole, kind of like water enters a drain, and it is gone. We don't see where it goes, but since it is thought that matter is the only stable energetic entity in the universe, we "assume" the black hole is made of dense matter. But what if the energy that enters the black hole forms a highly ordered galactic seed that is planted far far away? What If?

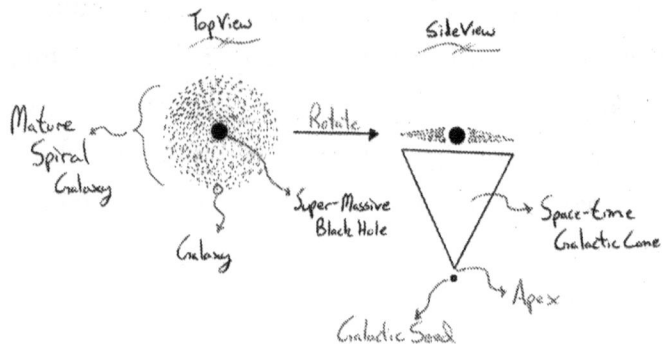

Figure 3: Space-time galactic cone created by a spiral galaxy.

If we imagine the Milky Way gave birth to an infant galaxy that is 500 million light years away, which is only about a 5% of the estimated size of the universe (http://imagine.gsfc.nasa.gov/docs/ask_astro/answers/971124x.html), then the light from that infant galaxy would take 500 million years to reach us. That is based off the assumption that an infant galaxy starts emitting light immediately. So if we think in galactic terms, even 10,000 light years away is far enough that we as humans, who have only been cognitive for about 5000 years, would never see the light from the Milky Way offspring. Therefore, it could be possible that the Milky Way is giving birth to new galaxies so far away that we'll never see where the Milky Way gave birth to a new galaxy. Personally, I'm under the assumption that most of us STILL think that everything that is in our space-time will only reside in

OUR space-time. I'm postulating that there might be an off chance that we are all connected to the greatness of it all! I mean, we HAVE to be connected to the greatness of it all for one simple reason: we are part of it. So if we look at a black hole as the birthing canal of our galaxy, which is giving birth to thousands and thousands of new galactic seeds that are millions of light years away, then this might explain a lot of things. Anyway, if our galaxy were to be an energetically closed system, then the more energy/light/matter/stars that entered the black hole would cause the black hole to grow in strength and size. This would cause a galaxy to collapse at a much faster rate. Obviously, this is not the case since galaxies last for a really really long time! Maybe someone should correlate the number of stars in a spiral galaxy to the size of the inherent super-massive black hole. Now Stephen Hawking identified that black holes emit radiation and postulated that they slowly dissipate over billions of years.(New Scientist, October 28, 2006, pg. 37, 5th paragraph) This would mean that all the energy/matter/light that has entered a black hole disappears. It is obvious that energy does not disappear, therefore it seems logical that the energy that enters the black hole is being placed somewhere else. But before we delve any deeper, let's think about how long a galaxy exists.

Does a galaxy exist for a million years or does it last for a couple billion years? Honestly scientists can only make an estimated guess and there is no general consensus. So let us just pick an arbitrary estimate that a galaxy lasts for 15 billion years (http://www.universetoday.com/am/publish/youngest_galaxy_found.html?1122004). Now if we think about how long humans have been paying attention to the cosmos with good telescopes, about 50 years, then we have been paying attention to about [(50/15000000000)*100] = 0.00000033 % of a galaxy's life. If we correlate that to something that is a little easier to understand, like a human life, then maybe you'll really understand how long we have been observing the cosmos. So lets say that someone lives a 100 years. Lucky them! Now if we were to look at 0.00000033 % of that person's life, then we would be looking at [(100 years*0.0000000033)*100] = 0.000033 % of a year. Since there are roughly 31536000 seconds in a year, then we would be looking at 10.4 seconds of that person's life. Now if we looked at 10.4 seconds of someone's life here on Earth, could we accurately guess how he or she got to where they are in their life? So why did I waste a whole paragraph to point out the relative time that we have been watching galaxies develop? Well, I just wanted to point out that there might be a SLIGHT possibility that our understanding of the universe and how galaxies form could be a little inaccurate, to say the least. First of all, astronomers have pieced all these theories together by looking at 0.00000033 % of the lives of a few selected galaxies. This would be like me looking at 10.4 seconds of your life, my life, your sister's life, and Abraham Lincoln's life to piece together one general theory on how life exists, evolves and formed. Heck, if we randomly looked at 10.4 second segments of each life, we still might not catch a glimpse of one of us copulating or your sister giving birth, especially if either happened in the darkness of space. Hopefully it is painfully obvious how little of the universe's development we have observed. This is why I feel obligated to toss some STUPID ideas around because EVEN IF we were to pay attention to the formation of galaxies for the next hundred years, instead of watching our favorite "reality" TV show, then we still wouldn't have enough information to make an EDUCATED guess on how galaxies come about. By the way, who the heck came up with this big bang theory anyway? Did they dream it up or did they actually see it? I don't think they saw it, so I guess that means they dreamed it up! Now I know what all the scientists are saying in regards to that statement, "The Big Bang theory is supported by empirical data that SUGGESTS that there was a BIG BANG." My reply to that is; "How often as scientists have we wrongfully come to a conclusion even when there is empirical evidence?" Honestly, most of the great discoveries on EARTH were done serendipitously. So I have some homework for anyone who has been paying attention. If you are bored with watching someone else's dreams on TV, go outside, look up, and dream about how it all began OR at least keep a look out for any asteroids that are heading towards Earth. Now I'll be the first to say my ideas are stupid, but if we ALL come up with some stupid ideas, SOMEONE is bound to "accidentally" come up with a GREAT idea! It happens all the time. It's called playing the Mental Lottery! It is important to remember that the Mental Lottery is like the regular lottery; you can't win unless you play! Just imagine if you came up with the next GREAT idea! You would be famous, you would be rich, and just think of the SEXY people you could pay to be attracted to your pocket protector. Now, there is a possibility that people will laugh at you, but a lot of great ideas were thought to be stupid at first:-) If you don't believe me on that, just read Bill Bryson's book "Earth* A short history of nearly everything." Anyway, even if society laughs at you, at least they're laughing and not throwing fruits and vegetables:-) Thus, it is my dream that everything is connected and something happened a long time ago that grew, gave birth, and now exists with me living on a very small part of its prodigy. Now here is a question to ponder if you feel so inclined; how many highly ordered spherical galactic seeds have survived?

The survival of a highly ordered spherical galactic seed

depends on the hostility of the environment in which the galactic seed was planted and the amount of energy that is inherent in the galactic seed. I wonder how much energy a galactic seed wastes on things it THINKS it needs? Here is a more realistic question; is the Milky Way giving birth to millions of galactic seeds? I would imagine that galaxies are like trees such that they produce thousands of seeds over their lifetime. I would imagine that galaxies produce billions of seeds because not every galactic seed will mature into a "stable" spiral galaxy. Now there are billions of examples in nature where an entity produces numerous seeds to ensure the survival of its species, but that's nature and probably has nothing to do with our galaxy. Or does it? Just think of how many sperm a male produces in hopes of fertilizing one egg to propagate the human species. Now that I've talked about how I dream of galaxies reproducing, it's time I start talking about how galaxies might be born.

Chapter Three
Blue light special on crazy spinning highly ordered galactic seeds, aisle 3.

N ow IF the Milky Way is giving birth to millions of galactic seeds via the super massive black hole/birthing canal, then this might explain why galaxies are moving apart at different rates, i.e., relativity! So if galaxy I produces a galactic seed, galaxy II, which is moving away from the galaxy I, then these two galaxies are moving relative to each other. Let just say they are moving away from each other at 5 m/s. (Figure 4)

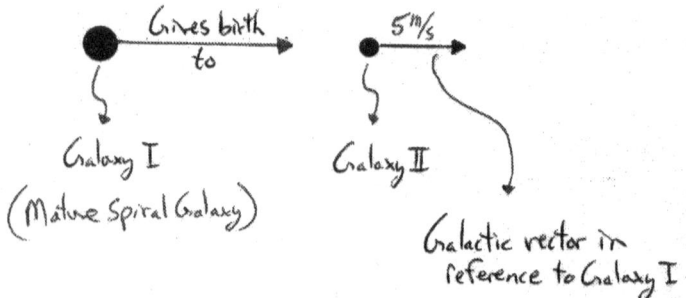

Figure 4: Galaxy I gives birth to galaxy II.

Now when galaxy II matures and gives birth to galaxy III, then these two galaxies are moving relative to each other, 5 m/s. (Figure 5)

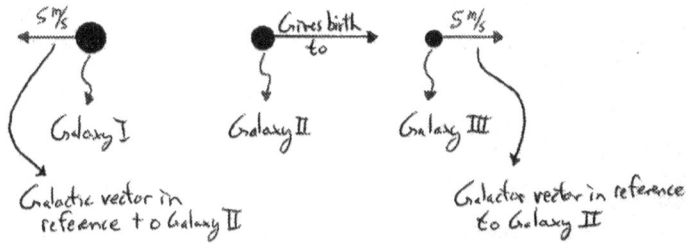

Figure 5: Galaxy II matures into a spiral galaxy and gives birth to galaxy III.

This means that galaxy III is moving away from galaxy I twice as fast as galaxy II is moving away from galaxy I. (Figure 6)

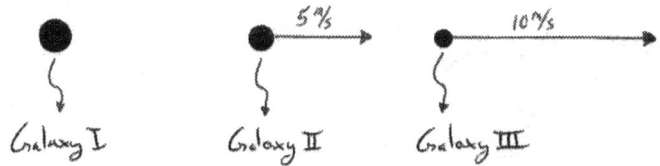

Figure 6: Galactic vectors in reference to galaxy I.

Yeah, I know it is stupid idea, but if we find the relative speed at which each galaxy is moving in reference to the Milky Way, then we could draw a GALACTIC family tree!!! Ok, THAT is just stupid!!! Therefore I'll start talking about the birth of a highly ordered galactic seed.

The birth of a highly ordered galactic seed from a spinning spiral galaxy might explain why spiral galaxies show similar spinning patterns.

If you are imagining the Milky Way is giving birth to an infant highly ordered galactic seed from the end of the super massive black hole that is at the center of our galaxy, then you need to remember that the stars in the Milky Way are spinning around the black hole. This would mean that if you draw a cone in which the base spans the diameter of the Milky Way, then the apex of the cone would roughly estimate the focal point of the super-massive black hole. Thus, since the Milky Way is spinning, then the apex of the cone is spinning much faster. (Figure 7)

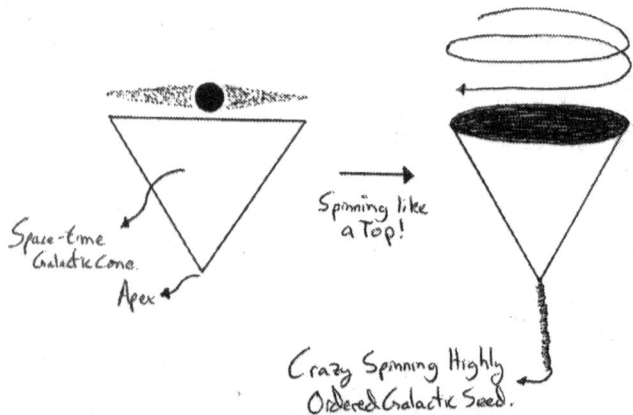

Figure 7: A spinning spiral galaxy produces a crazy spinning highly ordered galactic seed.

It is kind of like how the gears on Lance Armstrong's bicycle work. He peddles a larger sprocket that translates its spinning energy via a chain to a smaller sprocket, which causes the smaller sprocket to spin much faster than the larger sprocket. So if the Milky Way is propagating via the super massive black hole, then the galactic seed that is being born at the end of the super massive black hole (i.e., the apex of the cone) is spinning CRAZY fast.

Now it is common belief that the energy that enters the super-massive black hole is thought to reside as "super-dense matter" and creates a dimple in space-time, i.e., produces gravity. Personally, I don't think this is logical. If neither light nor energy can escape a place where space-time is bent, then this energy is probably following a certain path of space-time. I think the energy that enters the black hole is being decelerated. Granted, this sounds stupid since it is thought that matter is accelerated upon entering a black hole. I initially thought that too, but then I drew it out. It makes a little more sense that the matter/light that enters a black hole is being decelerated relative to un-bent space-time. Now you are really thinking I'm crazy, but hear me out before you put down this short book. I realize that it seems kind of stupid, but if the time it takes to travel a certain distance in our space-time is 100 seconds and the distance is 100 meters, then the speed is 1 m/s. Now if an object is moving 1 m/s along un-bent space-time and it hits a bend which causes the entity to travel a longer distance (500 meters) along that bent space-time, then it will take the object moving at the **same speed** a longer time to reach a **point** that is relative to un-bent space-time. Therefore the object is slowing down relative to the speed of objects that are moving in un-bent space-time. (Figure 8)

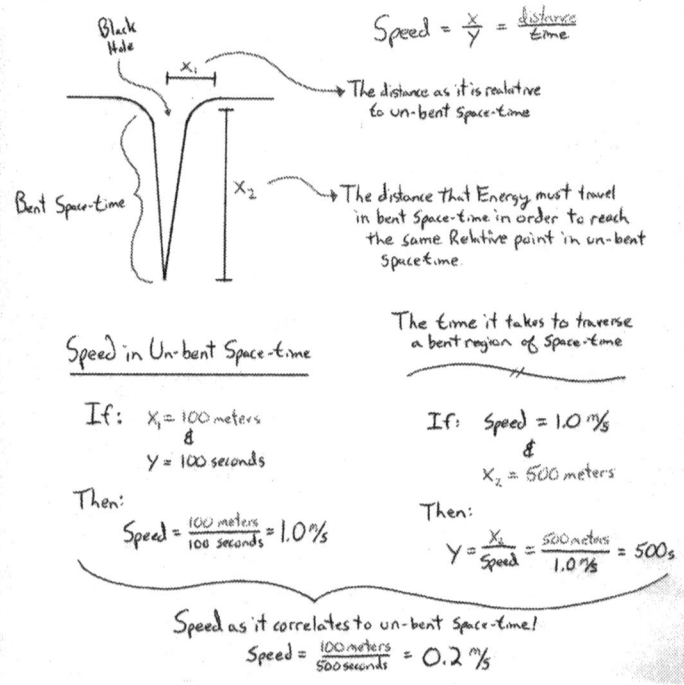

Figure 8: How energy is decelerated by bends in space-time.

Now I believe that everything I write is wrong because of my lack of empirical memorizable knowledge, so I will often come up with additional stupid ideas that explain the way I dream of things. My ONLY hope is that my stupid ideas will spark the truly intelligent individuals in science to view the universe in a slightly different way. These are only dreams, but they are the only things I have to hold on to right now. With that said, my next stupid idea is that the super massive black hole is a region of decreased "matter" and therefore a region of stretched space-time.

If you listen really hard, you can hear a million angry physicists shouting "Blasphemy!!!" Actually, I don't know why a physicist would use a religious term, but it sounded good when I wrote it. The reason that physicists are disturbed by this idea is because it is thought that black holes are made of super-dense matter. Now the reason it was POSTULATED that black holes are made of super-dense matter is because SOMETHING must be making this large magnetic moment that is responsible for the gravity that holds spiral galaxies together. What I don't understand this; IF charges spinning around a fix point in a field can create a magnetic moment, then why don't stars which are REALLY BIG CHARGES spinning around a fixed point, i.e., black hole, create a really BIG magnetic moment? So blasphemy away my friends, my stupid idea is that the super massive black hole is the result of all these HUGE CHARGED STARS spinning around a fixed point to create a LARGE magnetic moment. Yes I know it is simplistic and stupid, but I never understood why it hasn't occurred to anyone else.

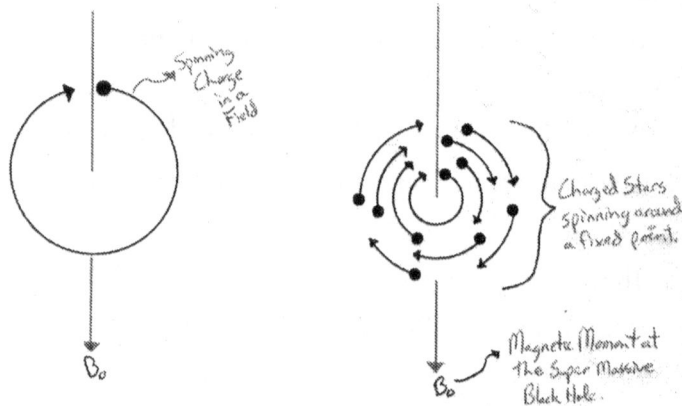

Figure 9: Galactic magnetic moment created by spinning charged stars.

So if we imagine that a black hole's magnetic moment is created by charged stars spinning around a fix point, then the accumulation of all these magnetic moments either act alone or in combination to form a massive magnetic moment. Either way, if you could correlate the number of stars in a spiral galaxy to the size of the black hole, then maybe the spinning star idea might be right. (Side note: there might be an entity made of matter that has a gravitational pull such that it appears dark, but why doesn't ONE fission reaction cause a chain of fission reactions that results in the entity to blowing up?) Anyway, the propagation of life idea and the fact that we are in a spiral galaxy leads me to believe that galaxies reproduce too. Yeah it is pretty poopie reasoning on my part, but I told you that I wasn't that smart;-) So the next question I would like to ponder is; how do all these magnetic moments combine within the black hole? Personally, I think it makes more sense that the magnetic moments are "coupling" together to create a larger magnetic moment. Therefore, the large magnetic moment at the center of spiral galaxies might be explained by the idea of linear superposition. For those of you not familiar with the idea of linear superposition, it is the concept that two things working together in a certain "pattern" will result in something greater than either entity acting alone.

Figure 10: Magnetic moment linear super-position produces one large space-time swirlie.

Now the pattern needed to cause linear superposition of the magnetic moments might take a little trial and error to determine, but I have no doubt that a brilliant physicist or someone that got bored with watching other people's dreams on TV will figure it out. It is possible that the magnetic moments created by each star forms an individual space-time swirlie, but who cares since nobody is taking the time to think about things greater than when to set the TiVo. So I'm going to start talking about the birth of a highly ordered galactic seed.

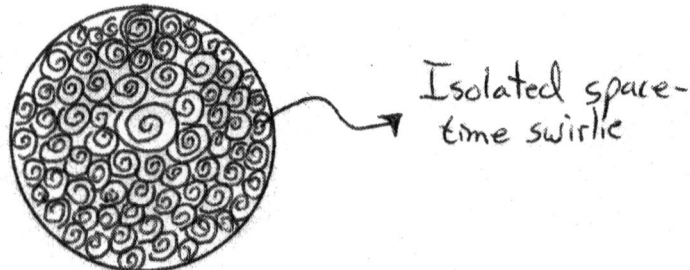

Figure 11: Isolated space-time swirlies.

Chapter Four

Awwww. Isn't that just the cutest little crazy spinning highly ordered galactic seed you've ever seen?

Now that I have proposed stupid ideas on how charged STARS spinning around a fixed point creates a gigantic magnetic moment that bends space-time to create a black hole/galactic birthing canal in which matter/light/energy might be decelerated to the point at which it breaks down into its energy components that are packed into a crazy spinning highly ordered galactic seed such that a galaxy might reproduce, I guess it is about time that I come up with a stupid idea on how a galactic seed grows into a mature stable spiral galaxy. As for what a highly ordered galactic seed "is" and how it breaks through into another space-time, I think someone needs to figure out how a proton is put together before someone comes up with those stupid ideas!

Once the highly ordered galactic seed has been planted

into another space and time, it diffuses in a circular fashion across that space and time as a result of the angular momentum of the crazy spinning highly ordered galactic seed. Once the energy distributes across a finite amount of space like a lightening storm to produce patches of highly ordered energy, i.e., stars, within the galactic disk, then the place where the highly ordered galactic seed entered that space and time is going to be dark and void of energy.

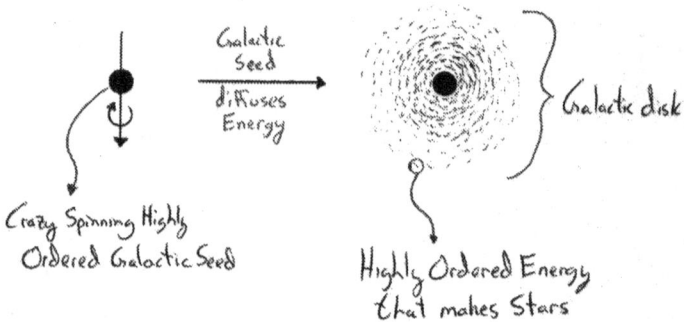

Figure 12: Diffusion of a highly ordered galactic seed.

Now you need to remember one thing about a galactic seed expanding into galaxy; the universe would be moving relative to you and the center of the galactic disk.

(Just a side note on finding life within our galaxy or other galaxies: Since our planet is made of heavier elements that are the remnants of dead stars, i.e., supernovas, then the planetary bodies which are stable enough to produce life will be at the outer regions of spiral galaxies.

First, the outer stars of the spiral galaxy are closer to the remnants of old stars, or as astronomers call them "asteroids." Second, the outer solar systems are more stable because there are fewer adjacent solar systems that will complicate gravitational patterns. Remember, one gravitational bump in a galaxy might cause a planetary body an enormous amount of stress. Now since we know that small amounts of gravitational stress may cause our planet to belch fire and brimstone, it would seem logical that a gravitationally STABLE system is needed to create a planet that is calm enough to produce life.)

Once a galactic seed enters a certain space-time and expands into a galactic disk that contains points of highly ordered energy, i.e., stars, the young stars start producing lots of space-time, i.e., matter. If there is any galactic potting soil, i.e., asteroids, that is within the region of the expanding galactic disk, then this matter will probably melt into magma that coalesces into large spherical planetary bodies and begins to orbit the nearest star. As these planetary bodies cool, all this EXCESS energy overcomes the activation energy needed to form a planet of complex molecules. As the star continuously pumps energy into the planet, the complex molecules again have the energy to form more complex arrangements, year after year after year after year for a couple billion years until a more ordered entity, i.e., organism, undergoes its first division. From there, the organisms spread like a wild fire by utilizing the chemical energy stored in the primordial soup until plants evolve. Then over billions and billions of years, plants transform the earth by directly harvesting the energy directly from the sun. Next, more and more complex organisms evolve until the Homo sapiens finally realized that they are alive. Now it **shouldn't** be blasphemy to say that evolution created this world of COMPLEX organisms because the Maker had to make us somehow. Instead, it should be blasphemous NOT to admit that an ENERGY GREATER THAN HUMANS created this galaxy, this planet, the plants, the animals, and YOU and I. Now I do not disagree that we have been made from an ENERGY greater than our imagination, but it is an obvious fact that there is a process by which things seem to be made. We have **slowly** learned this over the last thousand years. Now just because we have learned HOW things are made, doesn't mean we HAVE to give up OUR faith. Learning newer and greater things should make it obvious that WE don't KNOW EVERYTHING and we'll NEVER KNOW EVERYTHING. It was once thought that plagues were an act of the MAKER. Now however true that maybe, we NOW KNOW that bacteria and viruses cause plagues. Now it is a MIRICLE that so many scientists have spent their LIVES discovering these things such that a doctor can prescribe a medication that helps us live a longer and healthier life? So how can we repay all those people that have given their lives to make this world a better place? Sadly, most of them are dead but there are other people who are alive and in pain. Is there anyway we can help those who are alive? I wish I could have done more than just sitting around and dreaming up ways to help this world. I guess we all have dreams. Mine is that this world could be a better place.

I would imagine that the way the highly ordered energy distributes across a finite amount of space depends on the arrangement of the potting soil in that space-time. For example; if you were to plant a sunflower seed in some potting soil that is free any large rocks, then it will grow into a strait sunflower. Now if you were to plant a sunflower seed next to a cement sidewalk, then the cement will skew the growth of the sunflower. Therefore, if we correlate this to the expansion of a highly ordered galactic seed, then the crazy spinning highly ordered galactic seed's expansion pattern will be dependent on the big rocks, i.e., matter, that are in that space-time.

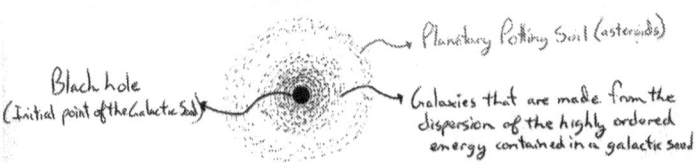

Figure 13: Galactic Disk (top view)

If there is a lot of matter in the area of the expanding highly ordered galactic seed, then that matter will absorb large amounts of energy and distort the expansion pattern of the galactic disk. The matter that resides within the expanding galactic disk will become super heated and will coalesce into large spheres to decrease surface tension, kind of like drops of water. These spheres will begin to cool and spin around the closet star. After millions of years of cooling down, the crust of the planetoid will harden. Only the planetoids that cool almost perfectly will form the conditions in which LIFE is able to grow. In our solar system, it happens to be the third planetoid. Then after millions of years of evolution and adaptation, we're finally cognitive enough to realize that we are here and what we do CAN make a difference. Personally, I have decided to waste my life by thinking up stupid ideas. It makes no sense to my friends, my family, or me. It is my way of coping with the poop that happens in this world. I might be crazy, but my craziness only extends as far as the pen and paper. So with all those stupid ideas under my belt, I'm going to come up with some stupid ideas on $E=mc^2$.

Chapter Five
E=mc² man!

L ets look at this equation slowly. Energy equals matter times speed of light squared. So if we had enough energy to accelerate a small amount of that energy to twice the speed of light, then we could create matter. Now the question I often wonder about is; "Why would energy want to make matter?" Well, I thought that fusion released lots of energy when matter is made, but why isn't there any factor to account for this release of energy in the equation $E=mc^2$? If $E=mc^2$ is the correct equation, then there is no energy released in the formation of matter, i.e., protons. So why do stars form matter and where does this energy come from that is used to piece together all these heavier elements? As you can see, something is missing in the current logic and I don't think the concept of gravity is going to pull all of this together, pun intended. Gravity is great, but I don't think it is the only force that dictates why everything happens. So from what I understand, which isn't that much, the energy inherent in a star is represented by the temperature of the star. So the first logical question is; where did the energy in stars come from?

I have heard that this energy comes from the beginning of it all, a big bang if you may, and that gravity somehow captured this energized gaseous matter ball, which subsequently ignited into a star. Once the star had depleted its energy supply, wherever that energy may have come from, then the temperature begins to drop. As the star cools down, this is when the heavier elements are formed. These ideas raise a couple questions in my head; if a ball of gas, which is already matter, forms a star, which makes matter, then where is the energy coming from to piece together all these wonderfully energetic and stable elements? If a star is made of hydrogen, then why do stars make hydrogen when they are young? If the energy inherent in each star is coming from fusion, then why isn't a young star making MORE heavier elements? In actuality, young stars are making of about 98% hydrogen (http://en.wikipedia.org/wiki/Sun), which is the same material that is "supposedly" collected together to make the star. Now it is known that stars make heavier elements through nuclear fusion, but this only happens as the star begins to cool and die. Therefore I propose that stars are made of an Energy Greater than Matter! Haha, I guess that's obvious. Anyway, matter takes up a lot of space and stars last for really really really long time. Therefore it would seem logical that if a star was a seed, then the star would pack as much energy as possible into a very small star seed. If you think about it, matter is too big and heavy to carry around the universe planting "star seeds." So the energy contained within a star is probably not stored as matter. Therefore, stars are probably made of Highly Ordered Energy or HOEs. Haha, that is only funny if you remember that men can be HOEs too;-) So the true source of a star's life is the energy that creates the heat inherent in each star. Now how does a star make all this heat and subsequently matter? Here is my stupid idea.

After a highly ordered galactic seed diffuses into a galactic disk in which the highly ordered energy has dispersed into smaller units, i.e., stars, the energy inherent in each star begins to lose order and form a cloud of energy. Once this cloud of energy, which is the key component of space-time, begins to bend then this is when the formation of matter picks up. Now the formation of matter truly begins when there is enough space-time to be bent such that the energy diffusing from the star can be accelerated to twice the speed of light, i.e., $E=mc^2$. Once the formation of matter steps up, this is when the star ignites. Now a star doesn't really ignite. What might happen is that all this matter, i.e., protons, that is formed by the star has to abruptly slow down. These protons must slow down because they are running into all the matter that is hanging around the star, i.e., FRICTION!!! Now many of you are saying "What the hell!? You can't accelerate energy to the speed of light, let alone to twice the speed of light because the acceleration of matter to near the speed of light will cause the matter to expand exponentially." All I have to say to that is, "Exactly!" If enough energy is being lost by the highly ordered entity such that it has to travel past a bent region of space-time, then the energy is getting accelerated to twice the speed of light as it is relative to un-bent space-time. Remember, it is all about relativity;-) Anyway, **Energy** must have been accelerated to twice the speed of light at SOME POINT in the formation of matter, otherwise $E=mc^2$ would be a useless equation.

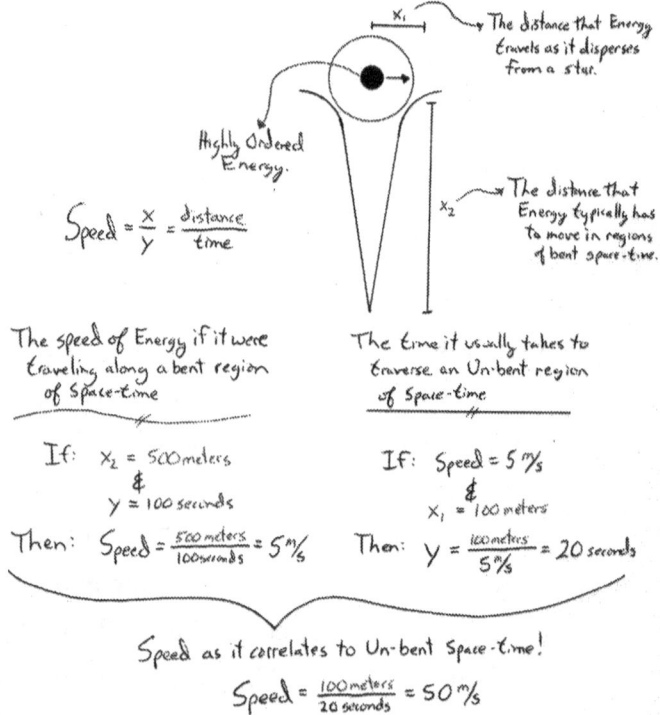

Figure 14: The Acceleration of highly ordered energy to twice the speed of light to form matter.

So when a star dies, the highly ordered energy becomes depleted. This means the star's internal energy, which is subsequently represented by temperature due to friction of the newly forming matter, decreases and the star begins to shrink. Once the star begins to shrink, nuclear fusion occurs to form heavier elements as a result of the pressure created by gravity. Now this is where things get a little tricky. Under the some conditions, the matter formed by the star becomes packed PAST the point that matter likes to be packed. This leads to one nuclear fission reaction that begins a chain of nuclear fission reactions and results in the explosion of the star. Under other conditions, the core of the star forms a supernova. I wonder what a supernova is? Maybe it is a stable fission star.

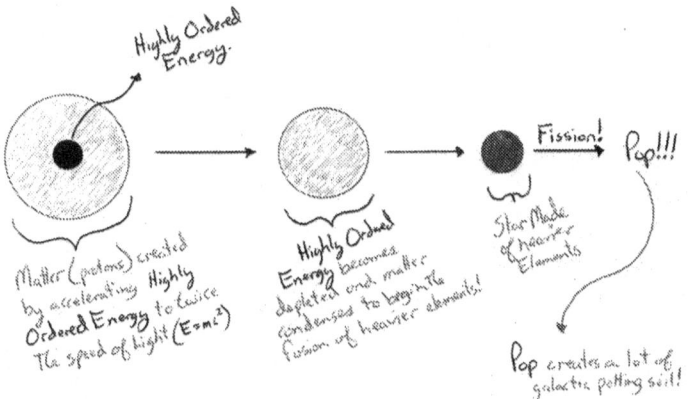

Figure 15: Death of a star.

One consequence of exploding stars is that they do

a great job of spreading galactic potting soil, i.e., heavier elements, all over the universe. Now even though a star might die within a galaxy, what causes a galaxy to die?

Chapter Six
Milky Way RIP.

The current idea of galactic death is that a galaxy dies when the black hole sucks in all the matter/energy from the galaxy and explodes again. Personally, I think that a galaxy started with a finite amount of energy, then as the stars die, i.e., loses order, their charges decrease. Now stars lose their charge by producing crap loads of space and time, i.e., matter, and this decreases the magnetic moment at the center of the spiral galaxy. Yeah, I know I'm stupid but please pay attention because this might save your life someday! Haha, just joking. As the charge of the star decreases, the magnetic moment that creates the black hole decreases and the galaxy slowly dissipates outward. Once the galactic magnetic moment is not large enough to hold the unborn highly ordered galactic seed in the black hole, the Highly Ordered Energy gets ejected back into OUR space-time. It will be a major bummer for the humans that MIGHT be around at that time, which I highly doubt, but they shouldn't fret! The highly ordered energy will probably cause a gigantic explosion that will send every solar system in the Milky Way hurling outwards into the depths of space to slowly die.

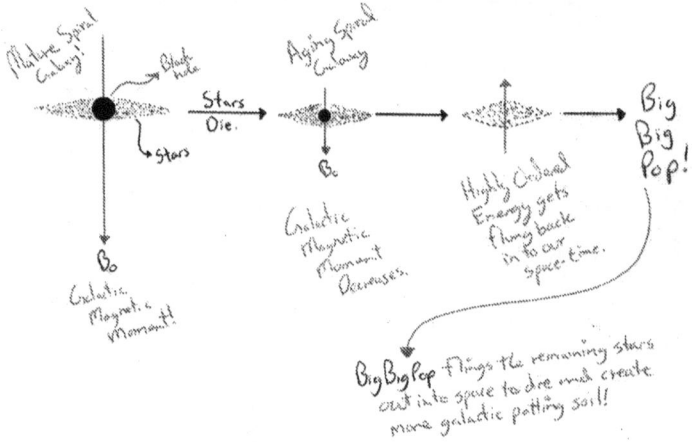

Figure 16: Death of a galaxy.

I know what you are thinking, "Why the hell did he say not to fret?" Well, it is as simple as the fact that the Sun in our solar system will someday die, explode and spread a large amount of planetary potting soil across the universe for the next galactic seed. Yeah, I know it is all about "ME" in our culture, but sadly it is not all about us. It is about HUMANS as a species. Someday another highly ordered entity would happen upon the remnants of our bodies and create another space and another time. Sadly though, we don't want to think about this because we still believe that it is all about us. When you think about humans in a galactic sense, we are selfish little monkeys throwing poop at each other because we don't care about the other living things on this planet. Yeah, I find it sad, but thankfully we will create a lot of planetary potting soil, after we make good old fashion Earthly potting soil. Maybe the next time a highly ordered entity occupies this space and time, we will be a little more considerate of the other people who have also received a chance to occupy this space and time on Earth. Yes, I know it is a dream that the world could be a better place, but it is not a dream that we are all potting soil, i.e., matter.

You see stars create space and time via the separation of positive and negative charges. This not only creates space via the separation of one thing, it also creates time via giving a specific integral of time that is defined by how long it takes for the positive and negative to find each other again. Now the Sun is a highly ordered entity that creates space-time, i.e., matter. Now I can't prove this, but it does seem logical to me. I think matter is made from highly ordered energy because it lasts a LONG TIME by simply separating the positive and negative charges. I guess this is the point where it becomes a question if there is any order.

Chapter Seven
Ordered or Not?

There are many people who think that the atomic nucleus is just a big mush of positive and negative charges. I would like to point out to those people that the atomic nucleus lasts a heck of a lot longer than humans. Now humans only last for about 100 years because our bodies can only keep ORDER in our genetic sequence for that time. The body has to keep our DNA safe. Our body also has to make the DNA accessible such that the proteins can be made that are allowing us to think, live, love and exist. I can only imagine how ordered the mito-chondrial process must be such that it is able to provide the energy we need to survive. Our bodies are the best example of how ENERGY must be organized in order to exist. The problem with the atomic nucleus is that it does not eat, sleep, or recharge in any way. The atomic nucleus also contains a lot of energy because it lasts for millions of years. When the atomic nucleus does fall apart, it falls a part with a big bang! Just imagine a nuclear bomb. So if you have something that has a lot of energy and it lasts for a long time, then we would call it a Battery. So I like to think of matter as a "Matter Battery" because it is such a PERFECT arrangement of energy that just keeps going and going and going without losing energy for millions of years! So the question comes down to which way is more economical for longevity; complete randomness or highly ordered?

We have seen in nature that it takes lots of energy to organize things. So it makes sense that the atomic nucleus is more ordered than the electron orbitals, but on a much smaller scale. If the atomic nucleus wasn't ordered, then it would have to waste part of its energy supply to organize the electrons into specific orbitals. Anyway, another of my stupid ideas is that the atomic nucleus is highly ordered entity because it has a finite amount of energy that will degrade according to the elements half-life. Therefore the organization of each element determines its lifetime.

Chapter Eight
What really matters?

My stupid idea on the atomic nucleus is based off the simple notion that negative and positive charges have a mutual attraction towards each other. I know that this is simple and stupid, but it might be the only thing that holds atoms together. Now I realize that my idea is stupid, so let's look at the current theory of the atomic nucleus. The current theory does not correlate the electron position around the nucleus as the result of any energetic interaction between the atomic nucleus and the electrons, i.e., columbic attraction. It is thought that a "strong force" holds the protons and neutrons together and the electrons just "happen" to fill these distinct orbitals as the result of some "unknown force." The current theory also uses the idea of a quantum jump to explain why electrons are never found in the atomic nucleus. Now I realize my idea is stupid, but let's imagine we are in the electron shoes. If electrons had shoes and WE were in them (size 5×10^{-42}), then we would be looking for what we are attracted to the most, i.e., the positive charge inside the atomic nucleus. Now our attraction to the nucleus is dependent on our distance from the nucleus, i.e., the closer we get to the positive charge the more attracted we are to the positive charge. Now there seems to be one problem with our infatuation with the positive charge: We CAN'T find the positive charge!!! Therefore, let us imagine we are electrons on the atomic orbital rollercoaster towards the atomic nucleus. Weeee!!!!!

If we are electrons that are heading towards the nucleus, what is keeping us from away from the positively charged protons? The current theory of the atomic nucleus has protons within the nucleus as "un-moving" entities that somehow keep all the electrons away from their positive charges. We know that electrons are moving really really fast and the 1s orbital electrons are the closest to the atomic nucleus. Therefore what is keeping the 1s electrons away from the positive charges in the atomic nucleus? Do the 1s electrons have a complex set of morals which restrains them from heading strait for the closest positive charge? Do they only like positive charges on the weekends? Do they have insecurities that cause them NOT to grab what they want and hold on for dear life? Thankfully, the brainless electrons only have one purpose in life: to find the positive charge! Electrons do not take lunch breaks, have breakdowns, or cry because they think they are fat. Electrons have been separated from their other half, the positive charge, and are as the Blues Brothers would say…"On a mission from God!" So what would happen if the electron found the positive charge? The physicists describe the reunion of an electron and a positron, the opposite of an electron, as complete **annihilation**! (http://en.wikipedia.org/wiki/Positron) SO WHAT IS KEEPING THE ELECTRONS FROM FINDING THE POSITIVE CHARGE?! Are electrons afraid of **annihilation** and therefore hide in their orbitals for millions of years? Maybe there is a miniature pit bull that keeps the pussycat electrons away from the positive charges in the nucleus.

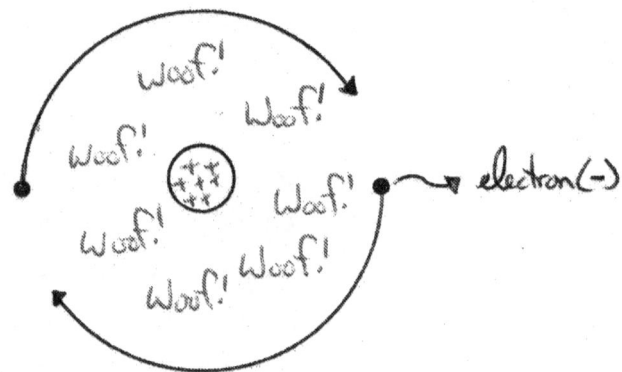

Figure 17: The miniature pit bulls that protect the nuclear charges.

Ok, the scientific community won't accept that. So what

if the electron is confused on where the positive charge is located in the atomic nucleus?

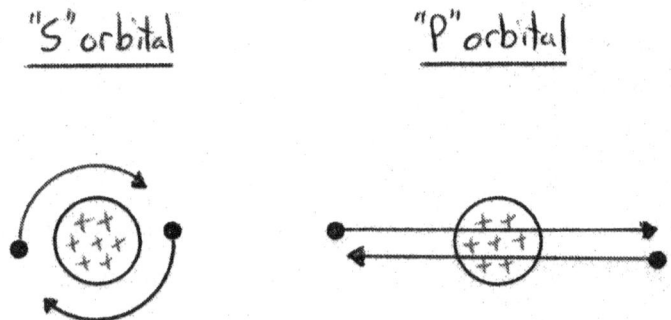

Figure 18: Elusive positive charge.

If the positive charge is moving, then this might confuse an EXCITED electron. Now the only problem with this idea is that the "current" model of the atomic nucleus states that the proton has the positive charge and the neutron is just a particle that sits around with its thumb up its atomic butt. If this is the case, then the neutron must be a reincarnated scientist, i.e., dork, which loves watching other charges get it on. The reason why localized charges within the atomic nucleus doesn't make any sense is because: 1) The electron can't find the positive charge. 2) The atomic nucleus contains a CRAP load of energy. 3) Something LOGICAL must be holding all these positive charges together within a small area. Now the "current" theory of the atomic nucleus says that there is a STRONG FORCE that holds the atomic nucleus together. Now it is obvious that there is a strong force, but what is causing this strong force? Is there a magical little gnome inside the neutron that casts a spell on the protons such that they stay in a certain area? Even better yet, neutrons are the Playboy Play Mates of the atomic particles and all the positive proton men are trying to get it on with Miss Neutron. So to date, a strong force is defined as a mystical force that has nothing to do with all the other observable forces in the universe. Well, since there is no magical force that creates this strong force, then what is really happening inside the atomic nucleus? I guess we should take a step back and look at how atomic nuclei are pieced together.

Chapter Nine
Hot Potato with a positron.

The current thought on atomic fusion is that two protons collide and there is a release of a Neutrino and Positron to yield the proton/neutron pair, which are held together by a strong force. (http://en.wikipedia.org/wiki/Nuclear_fusion) So IF two protons come together and release a positron and neutrino, then WHAT decides which proton releases the positron or neutrino? Does the proton **on top** get to keep control of the positive charge? Why do the two protons decide to become forever joined at the atomic hip by giving up the benefit of having two positive charges? Now the idea of sharing is not common in America, but what if a proton and neutron are sharing a positron? The idea of a proton and neutron sharing a positron might explain why a NUCLEAR force MUST occur on a NUCLEAR SCALE. This means that the two protons must get really really close before they decide to give up the benefits of having two separate charges and decide to share one positive charge.

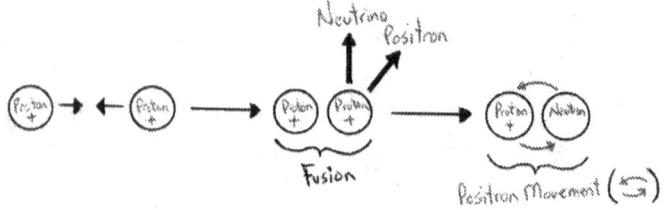

Figure 19: Hot potato with positrons.
(http://en.wikipedia.org/wiki/Nuclear_fusion)

Now this leads me to the next question: If a proton and neutron are sharing a positron, then how fast does the proton and neutron play hot potato with the positron? Since we can't see the atomic nucleus to pick out which entities are protons or neutrons, then the positron might be moving really really fast. Now how fast am I talking about? Since matter is dictated by Einstein's wonderful equation, $E=mc^2$, then my stupid idea would be that the positron is moving close to the speed of light. Now I think that a positron is moving at close to the speed of light because it is probably a harmonic of how fast the energy is moving inside protons and neutrons, i.e., twice the speed of light ($E=mc^2$). Yeah, I know I'm an idiot. Personally, I have only seen Einstein's equation used to describe the energy stored in matter and NOT as a description of how matter is made. Now I mentioned earlier how the highly ordered energy within stars get accelerated through bent regions in space-time to twice the speed of light to form protons, but you must not forget I'm an IDIOT. Anyway, Nerdy Neutrons can't just be placeholders that buffer proton interactions; they must be energetically interacting with protons. Therefore, IF protons and neutrons are SHARING a positron that is moving close to the speed of light, then whenever we blow this married couple apart, one of the entities will be holding the positron and will be forever labeled THE PROTON. So if a positron is exchanged between protons and neutrons, then a large cluster of protons and neutrons, i.e., the atomic nucleus, that are exchanging their respective positrons will create a pattern of positive charge movement about a spherical atomic nucleus, i.e., positive pattern.

If there is a positive pattern within the atomic nucleus, then it must be very stable because it loses a small amount of energy over an infinitesimal amount of time, at least relative to Humans. Thus the atomic nucleus, which is pushing and pulling these electrons into the orbitals around the nucleus a million times per minute, probably has a lot of energy and a ORDERED pattern. I know it is a stupid idea, but let's just imagine it for a moment. Now the pattern within the atomic nucleus must be more ordered than the atomic orbitals because a less ordered pattern cannot dictate a more ordered pattern without the input of energy. The last timed I checked, matter was not plugged into an energy source like the TV plugs into the outlet. So matter is energetically on its own!!! Therefore, it would seem logical that the atomic nucleus is more ordered than the atomic orbitals such that the atomic nucleus **does not** have to waste energy restricting the electrons into distinct atomic orbitals. For example: if an electron was a boomerang and you were the atomic nucleus, then you would have to throw the boomerang three thousand times a second into the same region of your backyard "atomic orbital." Ok, I'm exaggerating, but you still would have to throw the boomerang exactly the same way about a Ka-trillion times a week. Now for those of you are not familiar with Ka-trillion, it means A LOT! Either the atomic nucleus has killer biceps and a lot of EXCESS energy to throw the electron boomerang the same way every time OR the electron is just following a more energetic/stable pattern within the atomic nucleus. Personally, I think the electron is following a more energetic/stable pattern within the atomic nucleus, but I'm an idiot!

If the atomic nucleus isn't more ordered than the atomic orbitals, then the atomic nucleus would have to waste some of its **limited energy supply** to constrain the electrons into MORE ordered orbitals. If someone wants to postulate a way that atomic nucleus recharges such that it can waste energy to constrain electrons to defined atomic orbitals, please, be my guest. Personally, I'm convinced that the atomic nucleus is more ordered than the electron orbitals, but on a much smaller scale. Anyway, matter is like the greatest battery ever made. Sorry Energizer:-(Matter loses only a little bit of energy every couple millions of years before it decomposes into something else. So let's take a moment and consider how elements typically decompose and how this might have something to do with the positive pattern of the atomic nucleus.

The decomposition of elements typically results in an alpha particle, i.e., a fast moving helium atom of two protons and two neutrons. So is it a coincidence that a lot of elements will degrade by the loss of two protons and two neutrons or do two protons and two neutrons constitute the next building block of the atomic nucleus? So let us suppose that two protons and two neutrons make up the next building block of the atomic nucleus, i.e., two couples swinging. This would mean that multiples of four would yield more stable elements. So just by looking at the average molecular weights of elements, here are a few elements that are multiples of four: helium (4), carbon (12), oxygen (16), neon (20), magnesium (24), silica (28), sulfur (32), chlorine (36), argon (30), titanium (34), chromium (52), iron (56), and copper (64). Now oxygen, magnesium, silica, and iron are part of the eight most abundant elements on EARTH. The remaining elements mentioned above are KEY components of our bodies. Carbon, sulfur, chlorine, and copper are imperative to our existence. Maybe we are formed from them because their stabilities or maybe it was just RANDOM chance. It is your call;-) (http://www.windows.ucar.edu/tour/link=/earth/geology/crust_elements.html)

Chapter Ten
Where's the Keg?

My current stupid idea is that the proton and neutron are sharing a positron, which creates the STRONG FORCE. So I'm going to start from the beginning of the periodic table and come up with some stupid ideas for each element until I can no longer draw them accurately. From there, I'll let someone with a gigantic computer design the remaining atomic nuclear arrangements.

The electrons in the 1s orbital occupy the space around the nucleus such that they are as close as they can get to the positive charge and do not pass through the nucleus. Since the 1s electrons do not pass through the nucleus, they do not pick up any positive pattern, i.e., spinning patterned, and therefore will not create orbitals that are opposite spin. The movements of the electrons in the 1s orbital are spherical because they are following the movement of the positive charge around the atomic nucleus.

The electrons are symmetric about the nucleus because this will decrease the amount of repulsive forces between electrons. Now again, I must stress that the electrons are moving **much** slower than the positron, which is moving between protons and neutrons.

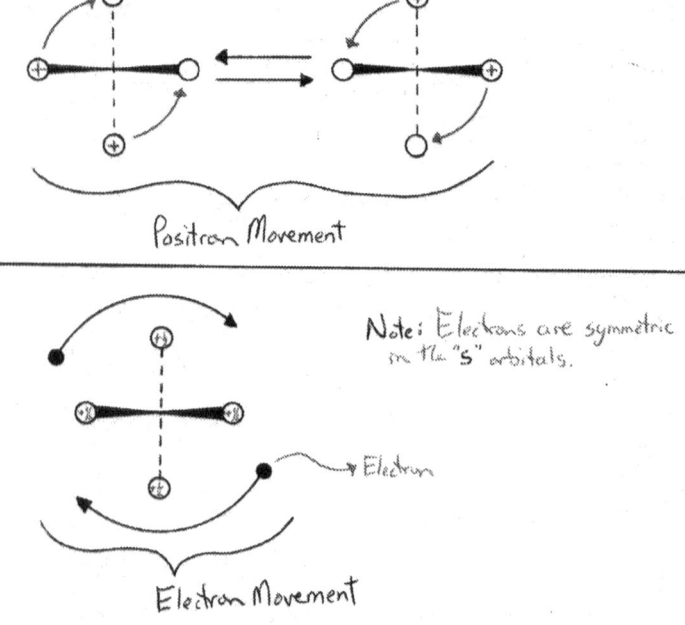

Figure 20: Movement of the electrons around the 1s atomic orbital.

The electrons follow the movement of the positive charge around the nucleus in a spherical pattern because the electrons can't find the positive charge. If the electrons could see the positive charge, then why don't they head

DIRECTLY for the positive charge? I guess the best way to explain this idea is to think of an electron as a thirsty college student.

Lets imagine an electron is a thirsty college student that is in search of the elusive kegger party because of the positive charge that is in the keg, i.e., beer. If the college student has the address of the kegger party and fairly good directions, then the trip will be a strait path for the kegger party.

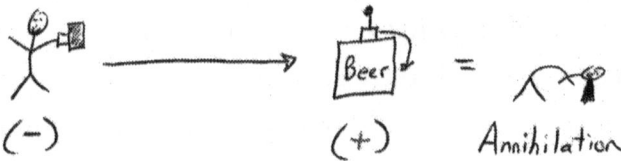

Figure 21: Address of the kegger is known.

Now if the college student has NO clue of the address and **only** has a general idea of the kegger's locale, then the college students will wonder around all night cussing up a storm.

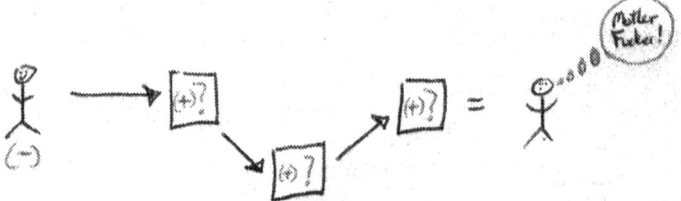

Figure 22: Address of the kegger is NOT known.

Now imagine that the kegger party keeps moving. Just think of how hard it would be to find the elusive positively charged container, i.e., keg. Now if we correlate this to the atomic nucleus, then this might explain why electrons can't find the positive charge. Now that I've described how the helium nucleus might be arranged, I'm going to move onto lithium.

I think that lithium's spatial arrangement within the atomic nucleus is as follows: one foursome (two protons and two neutrons) and a threesome. Typically, threesomes are not as stable as foursomes, i.e., the half-life of tritium is pretty short, but the threesome pattern is much more stable when included within larger atomic nuclei because the additional electrostatic interactions between adjacent proton-neutron pairs.

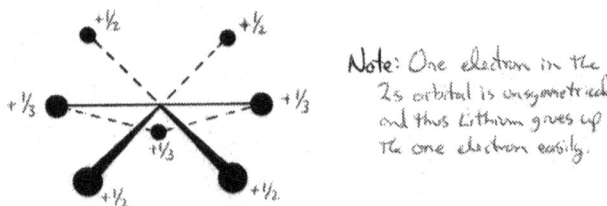

Figure 23: Arrangement of lithium's atomic nucleus.

Another reason why it is obvious that the "s" orbital electrons are symmetric is because one electron within the "s" orbital is NOT stable. Elements like lithium, sodium and potassium will give up that electron from the "s" orbital to become positively charged elements. Now I would imagine that the "p" and "d" orbitals are more accommodating to having one electron because the electron is passing through/near the atomic nucleus or is within a more stable orbital. (Note: I will only postulate stupid ideas for the major isotopes because the minor isotopes are probably just point variations within the more stable isotope such that a less stable rhythm is established, i.e., more repulsive forces which lead to quicker degradation of the element.) With that said, I will move onto beryllium after I talk about something that has been bothering me since I started dreaming about these stupid ideas on the atomic nucleus.

As an atomic nucleus gets larger, the stupid idea that protons and neutrons are exchanging a positron in a spherically orientated atomic nucleus becomes questionable because the distance between protons and neutrons is varying by a few picometers. Now many of you might be saying; "What the hell damon! You have already thrown traditional science to the wind and now you are worried about a Few Picometers?!" Now I understand your dissonance with my approach to these stupid ideas, but let me explain before you hunt me down and beat me with your wet and smelly gym socks. A few picometers on the atomic scale is like the difference between the orbit of Earth and Mars; important to life! If the nucleus was not designed in a spherical orientation with an empty center, then the size of atomic nucleus should not affect the stability. Also, if there isn't a pattern in the atomic nucleus, then the size of the element shouldn't have an effect on the stability and a cluster of protons and neutrons should have greater stability as more protons and neutrons are incorporated into the atomic nucleus. The obvious problem is that LARGER atomic nuclei become radioactive. The current theory doesn't describe why there is a "magic" size for atomic nuclei after which the larger elements start to decay at a faster rate. So what is causing these larger elements to be unstable? Personally, I think there are harmonics in the exchange of a positron between a proton and a neutron. There is also probably some vibration between the proton/neutron pairs as they exchange the positron. As an element becomes larger, the "core" proton/neutron clusters become stretched such that the positron must travel a greater distance. The further a positron has to travel between proton/neutron pairs, the more energy is contained between the two entities. It is also possible that larger elements constrain the proton/neutron pairs from vibrating as they exchange the positron. This would elevate the amount of energy contained in the element by decreasing another degree of freedom. Now it would seem logical that each **row** on the periodic table would dictate another harmonic between of the positron's movement between the protons and neutrons. It would also seem logical that each **column** would represent a given positive charge pattern within the atomic nucleus such that it dictates the element's electronics/properties. Ok, that was stupid and made no sense what-so-ever! I'll go back to my other stupid ideas.

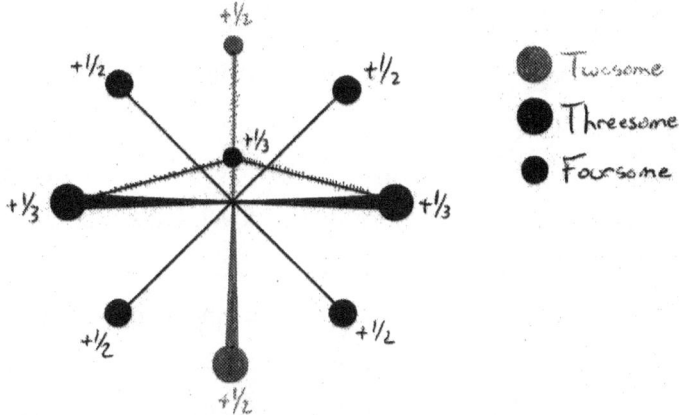

Figure 24: Arrangement of beryllium's atomic nucleus.

Now beryllium is a weird element. It would seem that the internal pattern should include a foursome (two protons and two neutrons), a threesome (two neutrons and one proton), and a twosome (a proton and neutron). Now it is apparent that I have attempted to draw lithium and beryllium as spherically symmetric, but it is obvious that the arrangement of protons and neutrons in the atomic nucleus will not be as spherically symmetric as I've drawn them. In actuality, the reason that these elements exist as positively charged elements might be because of their lack of symmetry about the atomic nucleus. So let's move on to boron and the "p" orbitals.

Since the "p" orbitals are often symmetric and electrons are moving really fast, it is highly unlikely that these electrons are stopping for a burger and fries when they reach the end of the "p" orbital. Therefore, there must be ANOTHER type of energy imparted to the electron such that is not correlated to the speed of the electron. I postulate that the electron passes through the atomic nucleus in search of the ELUSIVE positive charge and the movement of the positive charge within the atomic nucleus causes the electron to spin. The spinning electron creates a magnetic moment that slowly degrades by procession as the electron flies through its atomic orbital. In this way, the speed of the electron does not vary too much and it's the spinning energy, i.e., magnetic moment of the electron, which is degrading as the electron flies through the atomic orbital. This explanation also fits what is already currently known about "p" orbitals. If there are two electrons in the "p" orbitals, then each electron has opposite magnetic moments, i.e., (+) or (-) spin.

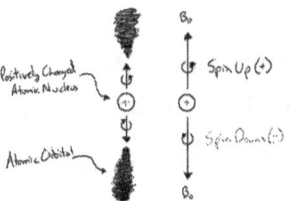

Figure 25: Spin up, spin down.

Now I can hear the professor's screams chastising me for even postulating that the electrons pass through the atomic nucleus. So let me slow down and make it completely clear that I'm an idiot. The current view of the "p" orbitals is that they are lobes of electron density and there is a node of no electron density at the atomic nucleus. Now it might be that the node around the atomic nucleus is the result of the electron speeding up as it approaches the atomic nucleus. As the electron gets closer to the positive charges, they speed up because of their attraction to all the positive charges within the nucleus. The further away the orbitals are from the atomic nucleus, the fatter and more spherical the lobes become as the result of the electron slowing down. As an organic chemist, we draw the "p" orbitals like an infinity sign. Now I don't know about you, but the "p" orbitals are **huge** arrows that are pointing to where the electrons are coming from and heading to. So many of you might be wondering why the two electrons in the same "p" orbital don't collide if they are symmetric and are passing through the atomic nucleus at the same time? If there is one electron in EACH lobe of the "p" orbital, then the two electrons in the "p" orbital are symmetric. If the two electrons are symmetric, then each "p" orbital should contain two paths through the atomic nucleus such that the paths are symmetric as well. This symmetry is important because it cuts down on the repulsive forces between the electrons. Now the shape of the atomic orbital might be determined by: the path of the electron taken through or near the atomic nucleus, the trajectory of the electron once its spinning pattern (magnetic moment) begins to degrade, the average distance the electron is from the atomic nucleus, as well as the other atomic orbitals (magnetic orbitals). Well, that's enough talking about the "p" orbitals. It is time to move on to the next element.

Since boron has five protons and six neutrons, then there might be two foursomes and one threesome. Now there is more symmetry in the boron nucleus which creates/stabilizes the "p" orbitals.

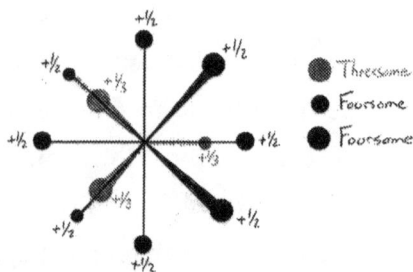

Figure 26: Arrangement of boron's atomic nucleus.

Now the arrangement of carbon might be three four-somes. Carbon is one of the most symmetric of atomic nuclei and therefore is one of the most stable elements. The symmetry within the atomic nucleus is apparent by its stability and its ability to symmetrically share its valence electrons.

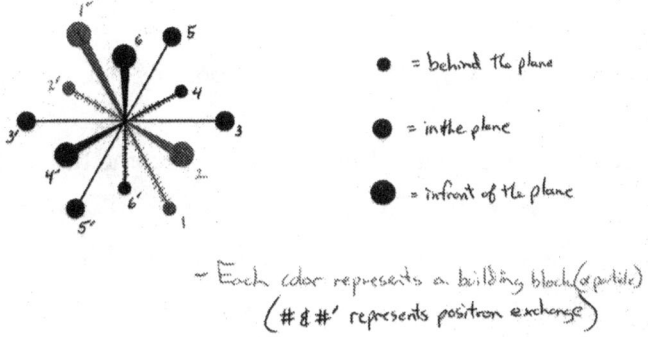

Figure 27: Arrangement of carbon's atomic nucleus.

Ok, I'm getting tired of talking about the patterns that no one else sees and after drawing carbon, I realized that my drawings are not doing justice to the arrangement of each atomic nucleus. Therefore I will stop here because I think you get my stupid point. I will make one last point though; as the atomic nuclei get larger in the third and fourth rows of the periodic table, there might exist a pattern around the midpoint plane of each atomic nucleus that incorporates an extended grouping of protons and neutrons past the stable foursome.

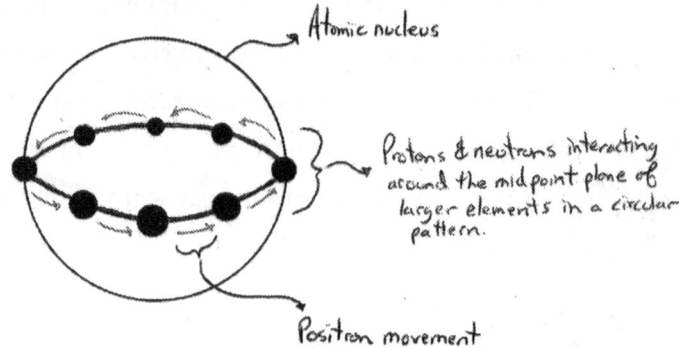

Figure 28: Extended grouping of protons and neutrons.

And so my saga ends. My ideas have become too stupid for my own mind. Therefore I relinquish all my stupid dreams upon the world because I give up. I'll never be more than a dreamer and all my ideas have been nothing but a waste of paper. I beg that the scientific community forgives me for blathering such BS. I went through hell in 2006 and these dreams were the only way I could escape all the pain in my head. Please forgive me.

Chapter Eleven
Just another stupid idea.

Now as scientists, we are trained to look for ways to test what we have postulated. So here is a way that you might test if black holes are decelerating matter. If the black hole is able to decelerate energy/matter to the point at which the energetic components of matter are released, then I have another stupid idea. By taking numerous magnetic moments and pointing them away from one spherical area, then you will be able to stretch a spherical area of space-time. By stretching that space-time, one might be able to slow down any matter that is trapped inside that space-time. Now the apparatus should be under vacuum to alleviate any unwanted molecules from getting trapped in this spherical area. Next, a stream of neutrons should be shot into the sphere of the stretched space-time such that one or many neutrons get trapped. Actually, this might be a good idea for fusion too;-) Finally, a pulse of energy should be applied to the magnetic moments such that a spherical area of space-time gets stretched. This might cause the matter in this spherical region to slow down.

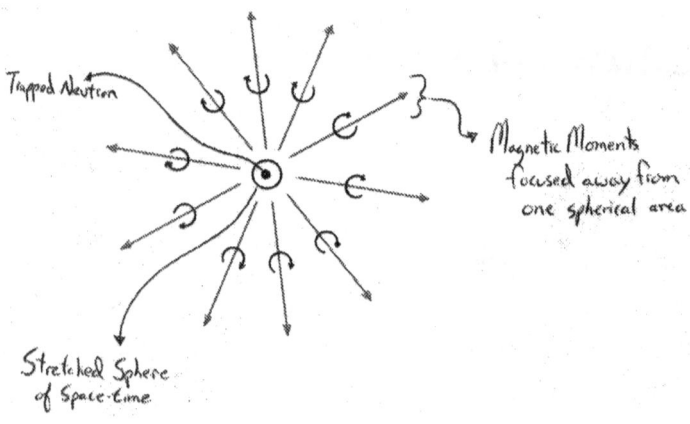

Figure 29: Energy source of the future! YEAH!!!!

By slowing down trapping matter in a region of stretched space-time, then you can manipulate that matter. One way to manipulate the matter is to slow it down and then turning off the magnetic moments so that the matter speeds up again. This is something that matter is not accustom to and if it is done right, then the matter might fall apart into usable energy. BAM! The ENERGY crisis of the world has just been solved! I wish it were that simple. Just imagine the schools we could build, the lives we could change, a world we could help. AMERICA could be a kind and compassionate nation that shares the riches of its knowledge with the rest of the world. Sadly though, someone will patent it and become so grossly rich that they forget to help the PEOPLE that have had a rougher life. If I came up with it, I hope I would have to courage to help society. I hope I would NOT become a rich bitch that needs to show off my money. Now I wouldn't mind not having to worry about money and maybe taking a vacation, but really, I don't need anything other than all these dreams that are right inside my head. So if this experiment works, then this might support my postulate that galaxies are made from highly ordered energy. My reasoning that a galaxy is made from highly ordered energy is as follows: 1) Matter is just too bulky to pack into a galactic seed (just imagine how much matter you would have to pack to start a galaxy. WOWIE! That would be a lot of matter partner!). 2) Stars must accelerate energy to twice the speed of light to form matter; $E=mc^2$!!! (Also, the energy used to form matter MUST be highly ordered because it loses energy very very slowly over millions of years!). 3) Space-time is relative (therefore entities can be accelerated or decelerated past what is normally thought in science as a result of energy moving through bent regions in space-time.).

Chapter Twelve
Going around in circles.

O k, that last idea was just a warm up stupid idea! Here comes the double whammy of STUPIDITY!!! Since light has a nominal amount of mass and charge, then it should be possible to collect enough light in one place without losing much energy to heat such that it could be used to induce a current in a wire coil. Whammy you've got energy to heat your TV or turn on your house;-) Haha, just seeing if you're paying attention! So how could this be accomplished? Here is my stupid idea.

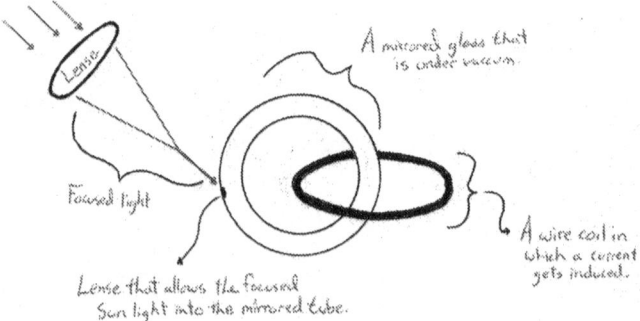

Figure 30: Going around in circles.

Wala!!! That was the stupidest idea in the World!!! I'm soooooo stupid;-) Now there is one key thing to remember about this stupid idea; the entry point of the focused light will have many variables that may affect this experiment.

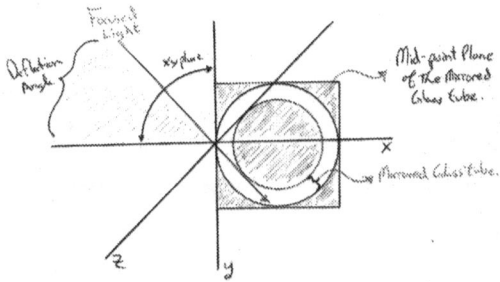

Figure 31: Deflection angle.

If a "y" axis exists on the midpoint plane of the mirrored tube, then the light entering along the "x" axis will reflect directly back out of the tube. Therefore, the angle of the focused light should be past the deflection angle on the "xy" plane. The deflection angle is defined as the angle at which the "initial" reflection within the mirrored tube occurs on the outer wall of the tube.

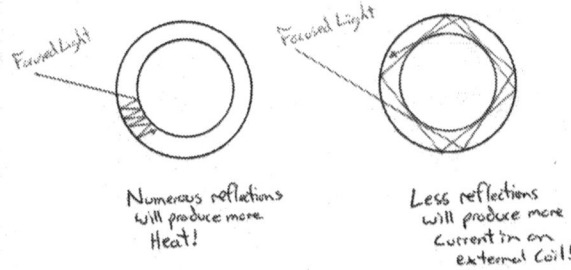

Figure 32: Reflection patterns within the mirrored glass tube.

By optimizing the deflection angle such that the number of reflections is minimized, then most of the accumulated light will act as a spinning charge that will induce a current in a wire coil. Now the movement of light out of the "xy" plane along the "z" axis will determine the how the light spins within the mirrored glass tube. The spinning of the focused light will be important, but remember that any movement out of the "xy" plane along the "z" axis will increase the number of reflections and thus additional points where energy will be lost as heat. It is possible that the focused light should be moving slightly to decrease the intensity of the light along certain points in the mirrored tube. If this is the case, I would suggest moving the focused light in a circle or spinning the lense, or both;-)

Chapter Thirteen
Why do bad things happen to Good people?

Why did the person you married turn out to be a pathological liar, among other things? This question will really mess with your head, but if you reflect on how you might have done a similar wrong to an undeserving person, then the question becomes; "Why did I hurt a good person and why didn't I do anything to make it right?" Did I think that my apology would not help? Was I afraid that my words would only make it worse? Am I still ashamed of the way I acted, for whatever reason, and I cannot bare the humiliation? Will the person not understand the reasons/circumstances that lead me to act the way I did? Why should I apologize when no one else apologizes? I'm sure there are a million more questions that might jumble our minds to the point that it is hard to think of anything other than pure hatred for the people in this world. Now many of your might be wondering how I would react if my ex-wife walked up to me and apologized for everything? Honestly, I have played this scenario out in my head a million times and each one ends with my words hurting her like the daggers of her lies that hurt me. I imagine saying things so eloquently evil about her person, mind and body that her soul would weep! Now I realize that my words would only be self-serving and would not play ANY part in making this world a better place. In fact, my words would create a verbal billboard saying, "Why did I have to go through such a shitty thing in my life because I don't think I deserve what was dealt to me and I want retribution for the events that sometimes happen in LIFE!" Yes LIFE is a game and even though we try and spin life's dial almost perfectly to advance ourselves to where we would like to be, sometimes the forces of friction from all those accidental bad things we have done to other people will guide our fate to something that completely destroys our soul.

Now many people will say that it is truly our "decision" if our soul is destroyed. From personal experience though, a life of complete failures and people taking advantage of my kindness is a little more difficult to overcome. It is truly the obstacles in our minds that are impassable. When these accidental bad things befall our paths, climbing over them is a little more difficult than climbing Mount Everest on a clear day.

So I might never apologize in person for all the awful things I have done to good people, but I want them to know that I'm sorry that my selfishness has put them through their "own" personal hell. I hope that each person I hurt was able to find hope in society because we all can't be bad, can we? Now I guess the next logical question after apologizing for my selfishness is how do I go about forgiving those who scared my soul? So what is forgiveness anyway? Well, honestly, a "Lack" of forgiveness is a survival technique. The more you seem to forgive people, the more they seem to take advantage of you. So to some extent, this world will become an angrier and angrier place because people have to fight to survive. (I guess it doesn't help that violence is glorified on TV and the sensual act of love is banned, but what does TV have to do with society anyways?) You see, back in the old days when you had your family around you, FORGIVENESS was the predominant survival technique since it was so difficult to survive alone. Now days, anyone can survive alone and have no problem meeting new people. Therefore, forgiveness has NOT become the dominant survival technique in our culture. The only hope for our culture is that as we become a poorer and poorer nation, which is the result of our substandard educational system, that forgiveness will again become the predominant survival technique. Now even IF the forgiveness characteristic is still predominant and I'm just depressed, it is sad that the wealthy have become so GREEDY that the abundant poor people have lost hope. It has become a dog eat dog world of greed and there seems to be no room for kindness if you want to survive.

So in closing, I have come to the conclusion, no matter how depressing it is, that I should not feel bad about not forgiving those who have wronged me because forgiveness in no longer the best survival technique in OUR society. Therefore I will continue loathing those who have wronged me and if they ever apologize, Cluck'em!!! Sadly though, it just doesn't FEEL right. Is this the "only" way I can survive in this world?

Chapter Fourteen
Science and the Maker.

I imagine that a religious person's hands are quivering with hatred as they begin to read this chapter. I understand that they must think I'm an arrogant butthead, so I'm going to explain myself quickly before they hunt me down and blow my brains out in the name of whatever Maker they worship. I am writing this chapter because SCIENCE is NOT about disproving the existence of a Maker! Instead, SCIENCE is about learning of the wondrous complexity that the Maker has created!!! Everyday I go to work and learn one more thing about how everything is dependant on its environment. SCIENCE is about ADMIRING the wonders of everything that has been created, day in and day out!!! So here is why every religion MIGHT want to support this highly ordered/ growing universe idea: First, religions have been loosing momentum because science has elucidated the fact that we did stem from dirt and that the growth and maturation of what exists here on this planet took longer than a couple days. Second, the idea of a highly ordered growing universe allows room for people to point a finger in the general direction of "God, Allah, Buda, or Bob."

I find it amusing that everywhere we look, there is complexity. There is complexity in the way matter interacts with itself, how your body works, and how the galaxy works. Sadly though, science always labels something as random when the pattern cannot be identified. If scientists would start thinking of the greater levels of complexity that have to exist, then maybe we can dream up some new ideas while marveling at what the Maker has made! If this were to happen, then I guess things could be different. Science has always limited its view by skepticism because there are a few nutcases, i.e., me. With that said, I still think it is good to dream about the complexities that MUST exist in the world. Personally, I find it intellectually stimulating to toss around some STUPID ideas that may spark the NEXT generation's interest in research. Now there are thousands of geniuses in the world, but sadly about 99% of them are still watching "reality" TV! We are simple animals that rationalize the world by what we see, but we forget how limited our view can be in front of the TV. We think we see it all, but we were NOT designed to see it all. We often forget that we can only see about twenty square feet of this world, depending on where we are standing. This means we often use our imagination more than we realize. Sometimes we don't even realize what our mind, i.e., subconscious, is thinking about. We are dreadfully flawed and yet we think we are perfect. I guess this a good description of the human species; Humans are animals that think they know it all because life was ONLY created for them. Now I think we are special, but we are not THAT special because whatever designed us made sure we only live for about seventy years. See my point? If we were so damn special, then why didn't the MAKER create us to last a little bit longer on this wondrously beautiful universe thingie such that we could figure out what it is all about? Either Homo sapiens are just another creation OR we were designed to work together to make this world a BETTER place. So here is my point for all of this rambling; if scientists were able to dream about the greater levels of complexity there exists in this world, but remain working on the ideas that can be seen, then a lot of stupid ideas can be tossed around and all this dreaming might jump-start some more gray matter. I think that it would benefit society if everyone was to believe in a religion, but it is important that the religion supports ALL LIFE NOW and in the FUTURE. As to why we are here, I suggest you look up to the sky and ask that question yourself. Best of luck! It doesn't seem to clear to me, but hopefully it will be clear to SOMEONE someday IF WE WORK TOGETHER.

(As a side note: Why do some people THINK their religion is the "right" one. If you were the creator and you made a diverse population of people, then you might have to explain certain ideas differently to each sect of the people group. I'm sure the Maker realized that there are so many variables in this diverse world and that one set of rules, ideas, or explanations was not going to cut it. It is like explaining something to your children when their eyes tell you that they don't understand. You don't keep explaining something to a child that doesn't understand and you don't explain something EXACTLY the same to each child. You look into the child's eyes and you have to start talking until the child understands one concept, then you build on that ONE concept. You have to start explaining something that the child can relate to, like boot or bird or milk, before you can move onto religion or physics. What I am trying to say is that many people claim to be in direct contact with the Maker. Now these PEOPLE are just PEOPLE and they make me very hesitant to believe in anything they say, but I find it amazing that so many people will believe ANY charismatic PERSON who claims to talk to the Maker. This is a problem because we forget to think about what we are ACTUALLY DOING and how it might directly affect

THE WORLD. In a nutshell, this world will NEVER be populated by ONE religion because everyone has a different perspective. I know we need guidance in our lives and that religion provides this balance, but what makes you think that Maker wasn't smart enough to realize that humans NEED to be UNIQUE! Personally, I think the MAKER was SUPER SMART! So since humans NEED to be UNIQUE, the Maker probably made a couple religions with similar core beliefs. Now the problem with having a couple of religions with similar core beliefs is that PEOPLE also LOVE to be RIGHT, which subsequently leads to isolation and a lack of communication. Now I realize that it has taken us a couple thousand years to realize that not everybody is right, but why has it taken us this long to realize that to COEXIST on this planet we MUST except that there is NO RIGHT RELIGION, i.e., there are **numerous right religions!!!!!!!** The problem is that there are people walking around this planet professing that the Maker has talked to them, chosen them/their families, and that "their" religion is the ONLY religion. How silly is that? Why would you walk around a planet filled with people who are sensitive about being unique, i.e., their religion, professing that YOUR religion is the ONLY ONE?! We might end up wasting a perfectly good life on this planet simply by fighting over who has talked to the MAKER. If we were to concentrate on learning about all the wondrous things the Maker has made, then MAYBE we could make this planet a better place instead of fighting.)

Now I am doubtful of everything I thought I knew about the world, science, and religion because I thought my ex-wife had a heart of gold. Sadly, she turned out to be a pathological liar. Major bummer for me, but in the end it was my fault☹ I can only describe it as being mind fudged. It turned my life upside down, inside out, and stretched my sanity and trust in society so thin that I was afraid I might never be the same. Unfortunately, I will never be the same. I have had a hard time believing in anything, such as Life, because it is so complex that there will never be ONE answer that is suitable to everyone. Sadly, this does not help me recover trust in my decisions, in my views, in the world, and in Life. This is where my dreams became my life. I was miserable, the world seemed like a miserable place, and I just wanted to find a way to make THE WORLD a better place for me, my nieces and nephews, and hopefully someday for everyone that gets the privilege to live on this wonderful planet. I realize that it is a dangerous affair to try and redefine the way someone looks at the world because I know how hard that transition period can be. I felt as if I had NO clue about anything and I hated everyone. Personally, I only made it through those tough times by dreaming about how I could make this world a better place. With that said, if I don't try and make this planet a better place through understanding and acceptance of our differences, then I just feel like a monkey walking around dogging the poop that others throw.

Chapter Fifteen
URUnique!

One of the greatest computers is about to be destroyed and nobody will ever know what complex problems it could have solved. It could have contained the unique processing ability to solve almost anything, but the bullet has left the muzzle traveling 385 meters per second. Those who were lucky enough to have escaped death's grip say that your life flashes in front of your eyes before you die. But when you realize, **truly realize**, that the bullet is going to end it all and there is no coming back, then your life MIGHT flash before your eyes. Once you realize that there is no coming back for a second chance, I wonder what you will think about. I wonder if you'll think about what you were going to have for lunch, if there was anything that you could have done to prevent your early demise, or what could you have done to change the world? Personally, I would probably think about my family, about all the people that I loved in my life, my friends, and my cats. Then if I still had time, I would ponder what it was all about, if there is an afterlife, and if I did what I was suppose to do while I was on this planet. Yes I know that death is depressing, but it is something that is going to happen. Now I wish I had enough time to ponder more, but there is no stopping the bullet that is going to scramble my brains. Now many of you are thinking that I am depressed, and you would be CORRECT;-) I will only ponder blowing my brains out when I'm terminally ill. Actually, if I were terminally ill, I would record my death in slow motion and in high definition so that you could see the LOOK in my EYES as shards of my brain created a UNIQUE pattern in the air and then a UNIQUE pattern on the floor. Now why the hell would I want to do that? Good question. It might have something to do with how precious a MIND and a LIFE can be. Never mind, it was just another stupid idea. Please return to your previously scheduled TV program…..

Chapter Sixteen
Epilogue.

So I guess I should explain why a YOUNG scientist, like me, would write a NON-SCIENTIFIC book on his DREAMS that are NOT SUPPORTED by ANY empirical evidence. Firstly, I went through a rough time in 2006 and this was my way of piecing together my life from nothing. Secondly, I wanted to portray that science and religion might NOT be the two extremes of the cognitive spectrum. Now why would I want to make this point? I think that IF the religious realize that science is about learning of the WONDERS that have been made, then the religious could concentrate on science instead of alienating themselves from the rest of the world. Anyway, if there is a JUDGEMENT DAY, then it might look GRRRRRRRREAT on your Life Chart if you helped discover the solution to one of the World's pains; like AIDS, CANCER, OR DIABETES, instead of being part of the largest CHURCH in three states, giving a bag of food on Thanksgiving, or re-reading the SAME book over and over and over again. I think that if you found something that helped thousands of people, then this might be at the top of your Life Chart. I can imagine the Maker reading your Life Chart on judgment day..."Well (insert your name here), that is quite impressive that you worked diligently for twenty years to discover a renewable energy that made the world a better place by decreasing pollution, decreasing the number of wars, and helping millions of children. This is one of the best Life Charts yet! Here, have one of my home brewed afterlife beers!" And of course you might reply like Homer Simpson, "Mmmmm, BEER." Thirdly, if we inspire more people to study science, discover new and wonderful things, and HELP the world, then AMERICA could regain the respect of the world. As for me, I really need to stop taking life so seriously, stop over analyzing things, and do more reading instead of dreaming. PEACE OUT!!! The End.